Deep Sea Mysteries

Deep Sea Mysteries

Seraphina Blake

CONTENTS

Introduction to Deep Sea Exploration

The majority of the world's oceans remain unexplored. The deep seas, lying beyond the continental shelves, make up nearly two-thirds of the surface of Earth. As sea explorers continue to learn more about this immense, uncharted wilderness, with conditions deprived of light, food, and oxygen, it is with increasing fascination that creatures are discovered. This essay will cover several key areas exploring the deep sea, including its history and significance, its social and environmental relations, and the technology advancements that allow humans access to a once-inaccessible world.

Humanity has always sought to explore further than its immediate environment. A grand mobilization of deep sea expeditions to smaller and smaller areas began in the wake of World War II and intensified with the development of the deep sea submersible as well as deep-diving vehicles which could descend as pressure-tight capsules up to the bottom of the sea. There are many scientific reasons to explore the depths of our ocean. This largely untouched and remote ecosystem is home to a range of strange and curious organisms; some of which hold the potential solutions to diseases that blight human civilization. Sophisticated technology advances in the 21st century also enabled mineral exploration and commercial mining activities to be in the deep sea. The implications of deep seabed mining on oceanographers and society are of far-reaching significance. We are just beginning to understand the biophysical and ecological characteristics of hydrothermal vent systems, cold seeps, and related deep-sea environments.

History and Significance of Deep Sea Exploration

Exploring the oceans has an incredibly long history dating back thousands of years. Deep ocean exploration, however, is a relatively recent branch of science. This is likely due to the fact that the deep sea has been largely considered to be from the abyss with which some might be better acquainted up to the mysterious fathoms. This began to change when the first practical submarine was designed and ushered in the era of modern deep sea exploration. Since the mid-20th century, we have conducted various types of deep ocean exploration with scientists first traveling beneath the waves on research vessels. We have sent machines to map the seafloor and collect water, sediment, and video samples in order to study the oceans.

The need to explore the deep oceans and create and foster a deeper understanding of what goes on in the largest ecosystem on the planet has only recently garnered significant attention. This is in part thanks to major events such as extreme deep sea challenges like the Trieste diving to the Challenger Deep and Innovator, the first remotely operated vehicle wading into the depths. Deep sea exploration has also gained power recently with the announcement of the United Nation's Decade of Ocean Exploration, affording scientists access to a fleet of new platforms, an operational satellite network that can live video chat with humans in one of the oceans' most remote areas, and an exciting future with new submersibles currently in design. Today, deep sea expeditions are team-based events, bringing scientists, students, teachers, engineers, and crew from all around the United States and the world to one place in order to share knowledge of the deep ocean. Furthermore, these expeditions include real-time interactions among scientists engaging with students and the general public through written forms and live video. These interactions allow us to bring the ocean to the front doorstep of the audience, inspiring men and women from all corners of the globe to continue to explore and protect the deep sea. With each new discovery, expeditious inquiry, and lesson unearthed, deep sea exploration is paving the way for new and exciting directions for research in this highly dynamic environment.

Technological Advancements in Deep Sea Exploration

Exploration of the deep ocean has depended on technological advancements. This extreme environment, with high pressure, low temperature, and absence of light, has typically been difficult to reach by humans. However, technological advancements have made several areas of the deep ocean more accessible for exploration. A variety of tools, such as remotely operated vehicles (ROVs), autonomous underwater vehicles (AUVs), manned submersibles, and camera landers, have allowed exploration of the deep-sea environment. Furthermore, sonars, such as side-scan and multibeam sonars, sea floor mapping, sediment coring, and water column sampling have also enabled deep-sea exploration. Imaging of these deep-sea ecosystems and characterization of the animals and communities that populate them is still a nascent field—much remains to be discovered about how these animals colonize and survive in these harsh environments, as well as how they interact to form communities.

Technological advancements and collaborative research have revolutionized the field of deep-sea exploration. Specifically, the advancement of ROVs has been a key driver in deep-sea studies. Such advancements have allowed the permeance and ability to study, sample—taking into consideration methodologies that minimize damage—and conduct experiments in never before regions where access and working capabilities were unattainable. Integrated ROV systems are now capable of mapping the substrate, such as multibeam sonar, and imaging the fauna with real-time applications and adequate lighting, such as high-definition cameras. Scalars of fauna size and fauna abundance and distribution can also be provided by landers and AUVs, including sediment coring. The development of various ecosystems of global interest that survive on chemosynthetic energy originating from hydrothermal vents and cold seeps and the incredible potential for biomaterials from the deep sea are examples of intentional discoveries by deep-sea researchers.

The Unique Ecosystems of the Deep Sea

The deep sea occupies most of the ocean, yet only recently have scientists explored these habitats using cameras, submersibles, and sonar techniques. Life here can occur deep in the Earth's crust, like primitive, heat-loving Archaea rappelling up to five kilometers inside the Earth. Amazingly, new species continue to be discovered in deep-sea habitats—from golf-ball-sized pink octopuses (Muusoctopus johnsonianae) to mini-eyeless, highly-primitive catsharks (Heterodontus vincenti). Deep-sea creatures can exhibit some fantastically extreme differences compared to life on the surface, such as glow-in-the-dark fish, worms with bacteria that allow them to live off a food source of toxic, hot oil; and shrimp that have a complex visual system and can hold in bacteria as a feast or for thinning out waste!

Adaptations for life in the oceans at such depths are numerous, with resistance to increasingly hot hydrothermal vent fluids at six kilometers depth being the most recently discovered. There is a lot of weird, rarely tasty, and even dangerous stuff in the deep sea, but so far we know of fewer than 20 extinctions! The deep ocean is unlit and without photosynthesis, so primary food production of plankton and upwelling productivity sinks down to the deep. Life here has thus developed to subsist on scarce, sporadic food falling to this distant habitat. Creatures in the bathyal and abyssal depths can bloat to 10 times under the pressure loads, afford the energetic extravagance of slow movement, and the enlisted equipment for enormous resistance can involve tooth-like denticles, bony armor, venomous spines, and shocking electric organs—while still relying on the less-energy-demanding chemical senses in which the deep sea is rich.

Adaptations of Deep Sea Creatures

As humans, we have spent a great deal of time exploring and exploiting the land, terrestrial life, and even the near limits of the Earth's atmosphere. Depths within the deep ocean remain some of the last great frontiers of Earth yet to be explored or exploited. Besides being interesting to explore within a purely academic interest, the animals and ecology of the deep sea is becoming of great interest for conservation and economic exploitation. While study of life at extreme sites and times in the universe is beneficial and of profound scientific interest, life in the still somewhat unknown and still, to some extent, mysterious deep ocean has the advantage of having a local "laboratory" close at hand.

Adaptation, or the changes to an environment or a change in the environment that helps many generations of animals to survive and reproduce better, can occur at three main levels: 1) morphological (the physical "outside" shape of the animal), 2) physiological (the "inside" workings of the animal), and 3) behavioral (tendencies, desires, of an animal). These adaptations help us understand why the organisms down there are as wondrous as they are. For instance, deep-sea animals have the highest in the animal kingdom longevity compared to the body size, which pumps their body with antioxidant enzymes and other features that humans could potentially imitate that help

an organism (in the case of deep-sea animals) resist damage from biological cells caused by high pressure.

Bioluminescence and Deep Sea Life

Where darkness has created an intense lifelessness on Earth, deep
sea canyons are teeming with life, sparkling in a magical display of
color and light. Diverse, with vibrant, often haunting and beautiful
wildlife-driven scenes with the biological version of a strobe light,
signaling an orgy of romance in some areas or a giant, glowing lure
that lights other areas beyond the reach of the sun.

Evolution has led to dozens of groups releasing their algae's organic light for different purposes. "Most deep sea organisms likely
make combinations of green and blue light," RR Hessler, a marine
biologist, wrote in 1968. Some organisms, such as loose-jawed drag-
onfish, can shoot cherry light because they outline lights and remain
red. By the 1970s, scientists discovered that some organisms can also
make green light, reporting deep sea fishermen who noticed green
in the eyes of caught fish from large, heavy nets. Over the years, it
has been found that a chemical known as luciferin produces storms
and oxygen that combine with enzymes known as luciferase to produce light. There are probably many more animals that can produce
light than there are, but you don't meet. There are fish in the open
sea that have tentacles of bioluminescence, but scientists have never
seen them alive. Bioluminescence can signal sex, even if it attracts to
attract the attention of the most attractive fish, attracting the light
of its attention. It can be repellent for other animals to see some-
thing more about to eat them. Bioluminescing lantern fish use their
photophore to meet. We don't know what most deep sea animals
use their bioluminescence for. For many, it may have several possi-
bilities. For example, the Belgian fish puts it out. It lives in a strange,
surprising world that, for all its strangeness, is incredibly long-lived

and historically consistent. Deep sea life is shown to be visionary because its denizens cannot see it. A glitter is wasted, in the open sea, the luminous organs useful for the manufacture of signals and steering the way out of the mud and silt of the bottom of the sea, where one does not react. Bioluminescence is also ecologically useful. In isolated, pretty places in the dark, a subtle flash of blue light, thrown by some hard quantity or another, comes at night, signaling its arrival in the thick of the seductive and strangely beautiful world, a domain, again. Desiring, a few moments before distributing darkness, very specific and rare marine life, lower caves exit back on. Bioluminescence helps you to bring up from the darkside whatever else is buried there.

3

Deep Sea Geology and Geography

When exploring the deep ocean, we have several environmental differences from the coast where most research has been done. On the coasts, processes are primarily driven by the sun and the weather, but at depth in the ocean, these factors are far less important in controlling the physical environment. Instead of sunlight, food sources are organic matter that "rains" down from the upper ocean. This is more rapidly and consistently transported in areas of the ocean where there are strong currents. These currents can also have a great effect on the temperature of the water in places like the Gulf of Mexico, where the "Deepwater Horizon" spill occurred. The currents enter the Gulf from the Caribbean Sea and eventually travel up to the North Atlantic and help influence the weather in countries like England. Deep in the North Atlantic, these waters become colder. When water is cooled, it sinks because it becomes more dense. These cold waters are known as the "North Atlantic Deep Water" and are also found far below us in the Gulf of Mexico and around this mud volcano.

Geologically and geomorphologically, the deep ocean is "younger" than the continents or much of dorsal areas along tectonic plate boundaries. This is due to a process called subduction where one tectonic plate slides underneath another. Subduction is a recurring process along many coasts of the world. The Mariana Trench is known as the deepest place on our planet. If Mount Everest were placed into the Mariana Trench, it would still be covered by over two kilometers of seawater. In addition to the mountains and trenches, we also find many underwater volcanoes. Frequently, scientists will find hydrothermal vents near local areas of these volcanoes.

Underwater Volcanoes and Hydrothermal Vents

These structures, stretching hundreds of kilometers, occupy over a million square miles of sea floor. Of these, over 5,000 were believed to still be active. In these tectonically precarious regions, molten rock from the Earth's mantle rises to the ocean floor to form underwater volcanoes, or seamounts. They rise from the surrounding ocean floor in conical shapes and often take the form of long ridges. Being composed of hardened lava, rather than directly from erupted materials, makes them quite different from "classic" volcanoes that the general public is familiar with. The process centers around the continuity of the underlying forces and materials and the nature of the crust they breach. While more common on ocean ridges, seamounts or so-called "guyots", that have since eroded to be level with the ocean floor, can be found virtually everywhere on the seafloor. The largest volcano on Earth, Tamu Massif, is an underwater volcano that spans over 120,000 square miles. Even so, underwater volcanoes often go unnoticed to the geologically uninformed. The abundance of similar features can be deceiving, and the size of seamounts only becomes apparent when compared with the average "height" of the sea floor.

Good things are known to come in large packages, but in the case of seafloor geochemistry, the most surprising finds are minuscule in scale. Hydrothermal vents are scattered along the tectonically vulnerable mid-ocean ridge and in spreading subduction zones. Here ocean water is heated by the underlying heat caused by magma build-up and circulation and undergoes a chemical reaction with the surrounding rocks. This gives the water distinct geochemistry, high in minerals and heavy metals. When released onto the ocean floor many meters below sea level, the minerals precipitate from the water, causing a steady growth of chimneys and the "black smoker" and "white smoker" vents they release to form. Black smokers take their name from the plume of dark minerals emitted and release waters at temperatures over 400 degrees Celsius. White smokers form from the precipitation of lighter minerals, releasing water of around 100 to 200 degrees Celsius. With their biodiversity and importance to the carbon cycle, interest in hydrothermal vent ecosystems has swelled.

Mariana Trench: The Deepest Point on Earth

Mariana Trench, the world's greatest depression, is the vast expanse of Earth's unique topography dotted with valleys, mountain ranges, and the famed Andes trenches. However, none of the valleys around the world is as deep as the Mariana Grate, situated in the Pacific Ocean. While the average depth is recorded as 11,034 meters or 36,201 feet, the lowest point known as "The Challenger Deep" is measured to be 10,994 meters or 36,070 feet. This immense measurement makes it unique to the genetic depression of Earth's seas called the "Deep Ocean," which begins at depths ranging between 200 meters to 1,000 meters and extends up to 11,000 meters or the normal sediment-water interface. As a result, the sediments rise up to 75% of the solid area and make this part of the Earth.

The floor of the trench is primarily made up of areas known as "guyots" or "level top" islands. These islands are not designated as higher elevation features that are not part of the trench. Also, the low-lying islands are mutually connected to each other through the Benioff construction or a regular cross-section boundary formed through contact. The insulating section formed from the east side is above the east and beneath the west, and not only allows the movable area to extend in the sea but also enables the elastic sediment to move aside. Additionally, the trench is also spewing several mineral incandescent springs. Volcanic activities are the responsibility and the direct result of the mid-ocean mountain range, which is the process of spreading sea-floor mulches without the invention of the submerged wrenches. Hormones inorganic can also hinder the plumbing effect from unleashing the lava from the farthest release in the trench.

4

Marine Archaeology in the Deep Sea

Marine archaeology is slowly revealing the mysteries and treasures of our seas, making us imagine ancient customs and habits while capturing the changes that took place over time, as well as faster and abrupt changes. The history of shipwrecks is of great interest to the general public. It evokes the age-old urge to get close to secrets and treasures, the atmosphere of adventurous exploration, but it also tells the history of our naval heritage in the face of dangerous seas and the randomness of navigation. Shipwrecks - whether archaeological or accidental - are linked and inscribed in the richness of navigation, trade, power structures, conflicts, technical development, economic fluctuation, and social history. Where mining heritage inverts or reverses the on-land picture, turning consumed soil mines into lakes or land excavations into landfills, at sea wrecks and submerged activities are additive and offer valuable insight into lost things. Therefore, marine archaeology ends up being an unavoidably interesting area of study.

Shipwrecks are rich and exciting, and indeed tantalizing in terms of their underlying reality, as countless secrets, mysteries, and treasures lie "underwater." Whether encountered as desolate, deserted wrecks silently watching the seascapes underwater or seen as dynamic and grandiose, as living expressions of naval technology, control, and culture, shipwrecks are certainly entrancing. The present overview, of necessity, only encapsulates the vast and fascinating realm of underwater archaeology. It is merely an initiated view, a holistic approach at best. For those interested and who immerse themselves in the study, underwater archaeology is an opening into infinite and unexplored worlds, each of which calls for attention, investigation, and, admirably, understanding.

Shipwrecks and Lost Civilizations

Imagine a time traveler transported back hundreds or thousands of years. Since most of the world's population resided along coasts or major rivers and used these aqueous highways as their highways or interstates, chances are they would find either a shipwreck or a harbor or sailing ship within a few miles of their landing spot on any land mass. In no place on earth is this more graphically demonstrated than in the Marine Archaeology Section of the San Diego Museum of Man. The centuries and millennia have condensed the past to the point where cultures, both human and sub-human - and the landscapes they created - have blended into each other. The truly amazing aspect of marine archaeology is the goal. Ships have always been the primary movers of civilization since it took to the water, but unless one is exposed to that singular fact, the remaining 39,997 feet of water on top of the Corinthia is not usually a consideration to the general populace. And while actual discovery at sea gives one what some would aver is a 'scholar's high', as will be demonstrated in the brief section, it is more true that the identification of a new shipwreck or a lost, submerged 'civilization' almost never occurs when it is sunk or inundated. As such, marine archaeology functions as a superb example, prevailing over a hesitant audience and archaophobic local, of cultural history at its most exciting. Deep sea expeditions in the last decades to such locales as the Peruvian Coast and the Black Sea have led to the examination of tons of raw archaeological debris, and have opened up new avenues of cultural study. However, and most unfortunately, the academic community has been relatively slow to respond, and so only painting in general overviews of these deep water riches, leaving largely unscathed the masses of all-inclusive debris available for inspection.

For others, shipwrecks deliver human tragedy, technological advance, trade and commerce - not to mention the plain beauty of a vessel under canvas. Shipwrecks are the single embodiment of the faceless masses of the many 'stories' of deep time scholarship. The craft themselves, like medieval members of the clergy or the Greater Second World War mechanical power output of the United States, prove the Great Man hypothesis false. They give names attached to the featuring artifact types - names of battles, discovery, trade/mercantilism, and overtime, lifeways and cultural interaction which have drawn many to the same dog-eared textbooks whenever it comes to Europe of the Past. Even without artifacts, shipwrecks offer romanticism and failure, the trials of humanity in conquering the final frontier of the world even if it is only the Terran part suffice. For the shipwrecks that can be linked with certain navigational or design improvements, early writing, pictographs and the codification of bronzecraft prove momentarily fascinating to documentary filmmakers, the breathless equivalent to the viewers of Nova. Early modern exploration shipwrecks, such as the ill-fated San Felipe in the Great Water, sensually slam home exotic Novae, depicted through historical perspectives as High Adventure along the edges of the known/unknown world. This unique mixture of pitfalls and triumph also pertains to much of the work done at sea by students of the ocean's past. A flawed combination of detective work, diving, and ill-advised sun exposures that distort vision, this archaeological smorgasbord creates a long-suffering indulgence of silliness and memory loss.

Preservation Challenges and Techniques

The preservation and study of this material in itself is a deeply complex challenge. As it was inundated in sediments for thousands of years, these materials are no longer suited for display or shown to the public as is. Conservation is a complex and expensive process to make any metal, timber, bone, glass or ceramic object stable and durable. Just as with exploring in the air or on land, work in the ocean quickly encounters many significant challenges.

Marine archaeologists must be incredibly well-versed in archaeology and anthropology, but also highly skilled divers and scientists. An entire subfield of the bigger marine archaeology field is devoted to exploring and understanding these remnants of human history from the deep.

Special logistics, knowledge, and abilities are essential to be able to function there, as they are part of a working environment where breathing would be dangerous. And looking for and recording archaeological objects, documenting information, and taking pictures underwater brings the additional challenge of coping with the key ways of a world otherwise unknown to mankind.

Water throws combinations of currents and force around these locations, which render standard approaches inefficient. As a result, a number of innovative research methods such as remotely operated (ROV) and autonomous (AUV) are being employed to perform archaeological surveys and interventions away from the shore.

5

Environmental Threats and Conservation Efforts

Environmental Threats It's only in the past 150 years, however, that we have learned even the barest minimum about life in the deep sea. The massive resources and growing interest in exploiting the resources have outstripped our knowledge, and we are uncovering the staggering biodiversity of the deep sea and the value of its resources only as, and possibly after, technological advancements are enabling exploitation and extraction. Increasingly, human activities are affecting the deep sea, exacerbated by technological developments and ocean climate change, and they now pose serious threats.

Conservation In the face of these escalating threats, an increasing number of conservation strategies are being employed to protect communities and individual species. In 2006, the United Nations began discussions on how to conserve unique habitats in the deep sea. In 2008, the General Assembly agreed to develop a new international instrument to manage operations in areas beyond national jurisdiction through a series of intergovernmental consultations called the "Informal Working Group". Conservationists have challenged the detail of the decision, arguing that some marine areas had become reserves in name only because they lacked the resources and authority to manage these places. But they did not challenge the larger strategy that reserves are one way to protect wildlife and maintain natural systems over wide areas. The irony of these forward-thinking conservation strategies is that we still know so little of what we might conserve and how to go about it. The deep sea could be home to millions of species, many, possibly the majority, as yet undiscovered. Species living in these systems can have life histories, range sizes, and endemism that is extraordinary even in marine or terrestrial terms. The race to discover and inventory deep sea life before it is lost is thus of the highest urgency.

Deep Sea Mining and its Impact

Deep sea mining is an emerging activity that is mostly driven by an anticipated increase in the demand for minerals. These minerals are found as seabed massive sulphide deposits, polymetallic nodules, and polymetallic crusts over a depth of 1,000-6,500 meters in international waters or Exclusive Economic Zones (EEZs). Seabed deposits in international waters are mostly polymetallic nodules and are located in the Clarion-Clipperton Zone (CCZ) of the Pacific Ocean. Those in EEZs are mostly seabed massive sulphide deposits and are found in the Western Pacific, particularly in the waters around Japan, the Philippines, and Papua New Guinea.

Deep sea mining promises that the areas containing these (sometimes solid pelagic) deposits made up of different metals including gold, silver, copper, cobalt, lead, and zinc would be a new resource frontier. In addition to the estimated value of minerals ($455 billion in the CCZ), they are seen as promising sources of metals essential for the emerging low-carbon technology sector. One of the primary discussions on deep sea mining is about its possible impacts, especially the harm caused by vast habitat destruction (for polymetallic nodules and crust mining) or alteration (massive seabed sulphide deposits) in the deep-sea environment which is still largely unknown to humans. The economic success of deep sea mining, even if some stakeholder scenario models are correct and hype is to some extent substantiated, must also be doubted. Deep sea mining also raises a number of ethical, policy, and governance issues, one of the most prominent of which is environmental injustice. It is yet to be seen whether compensatory justice (providing more extensive economic benefits or just monetary gain) will do anything to mitigate the ethical dilemmas deep sea mining has agitated in a number of countries.

Conservation Strategies for Deep Sea Ecosystems

There are several conservation strategies implemented to protect deep-sea ecosystems and to promote new sustainable approaches to fisheries in international waters. Approaches proposed to protect the environment include: (1) protection of deep-water coral reefs (including those of other Lophelia coral species); (2) addressing, through closed areas, a range of longline and trawling opportunities and constraints; (3) habitat protection or management in relation to the conservation of the giant grenadier; and ecologically (4) description of species and habitats which could be designated as "Vulnerable Marine Ecosystems".

Habitat protection: There is an international and national interest in protecting deep-water coral reef habitats, particularly those with a presence of other species of Lophelia. The main approach which may involve the protection of these reef features is greater marine zoning, which could involve managing physical damage to these features from bottom trawling.

Species conservation: The giant grenadier may be an important component of some deep-water fish assemblages in the region. Elasmobranchs, specifically Greenland sharks Somniosus microcephalus and sleepers Dalatias licha, are also present in by-catches, and are taken off the east coast of New Zealand.

Marine protected areas: There are currently about nine voluntary marine protected areas which may be including below the 2000 m contour which may capture a degree of the existing benthic biodiversity of the high-seas expected to occur beyond that depth. Some management measures in place are assisting voluntary reduction in fishing effort in the main operational areas. The 10% of the EEZ bottom trawling closures introduced on the Chatham Rise were partially implemented. In the South Tasman Rise, the establishment of a High Seas Fisheries Strategy focusing on deepwater fishery management is proposed.

6

Future Prospects and Discoveries

Most of the ocean is completely unexplored and many new marine species could be living in the unexplored habitats of the deep. While it is estimated that new discoveries have been accumulating at a smooth rate over the past few decades, exploration of the deep sea has increased since the 1990s. Already, new species have been described and many new phenomena have been discovered. Especially the sparse exploration of the vast hadal zone (6,000 to 11,000 meters deep) and its treacherous habitats of trenches present many sites waiting for unmanned deep-sea landers to offer new direct oceanic records, yielding several new discoveries.

Also, it will be a matter of time for the submarine DSV Limiting Factor to be offered for commercial tourism into the hadal zone, causing an even greater interest in the world's most remote frontier of marine life. Furthermore, shortly, two new exploration tools will enhance the exploration of these deep trenches, i.e. the drones that could explore the abiotic geological characteristics of the ocean floor, and deep-towed camera systems are starting to explore hadal plains like the Peru and New Guinea trenches as well. Future technical developments will even add to the knowledge of the deep sea and mid-water oceans for the near future. Furthermore, a manned trip to the deepest point on Earth, the Mariana Trench, also holds great excitement, its biota, and new discoveries to be made. It might take until about 2030 when the Arctic Ocean becomes an ice-free (mid-oceanic) summer, and future discoveries can be combined from the ice-covered world of both poles.

Unexplored Regions and Potential Discoveries

Despite how technologically advanced our world has become, much of the ocean remains a mystery to us. There are several regions of the ocean that are relatively unexplored due to their lack of proximity to human settlements or their extreme depths; while we have a sense of their topography thanks to sonar readings, we still have little firsthand knowledge of what life can be found within them. As is the case with much of space exploration, there's no telling what breakthroughs and compelling discoveries can be made by studying these unexplored environments. While the depths of the ocean floor are largely not as unwelcoming as undiscovered planets or moons, they are still home to a rich variety of creatures that provide tantalizing mysteries to solve.

The allure of the unknown is a powerful source of attraction, and the deep sea - which is still in many places uncharted and largely misunderstood - holds an abundance of secrets that we are only beginning to unlock. The Hadal Zone is the region of the ocean floor found at water depths between 6,000 and 11,000 meters, which includes the Challenger Deep in the Mariana Trench - the deepest point on Earth. While the trench itself is the focus of both scientific study and casual public interest, most of the Hadal Zone remains similarly mysterious. It is believed to take up approximately 1% of the total volume of Earth's ocean, but less than 0.005% has been surveyed. Remarkably, it was only in 2020 that researchers discovered how the Hadal Zone is connected by deep-sea trenches.

Emerging Technologies in Deep Sea Exploration

The last 20 years have seen significant improvements in the technologies available to ocean scientists to explore regions that have never been visited before. The development of piloted submarines that can take explorers to the bottom of the ocean and remotely-operated vehicles that can gather samples have transformed deep ocean exploration. While the cost and access to these kinds of tools may still limit who can use them, other sampling technologies have inspired a new generation of scientists, who can now map, sample and explore the ocean from the comfort of their own labs. Imaging techniques have been revolutionized by the development of digital cameras, hyperspectral imaging and 3D reconstruction. Thanks to the precision and detail made possible by these, taxonomists can now work with spectacular natural light photographs of even the tiniest objects (including field-collected samples). The simultaneous development of small, even pocket-sized genomic analyzers is finally bringing high-quality DNA analysis out of the lab and into the field. There are very few limits ruling out the possibility of what can be done, this has been a game-changer for uncovering the secrets lurking in deep-sea sediments and waters.

One technology which particularly impresses me, thanks to my background in fine-sediment analysis, is the work of Ocean Networks Canada, it conveniently plugs into an existing ocean observatory on the floor of Barkley Canyon (off Vancouver, Canada). The experiment features a deep-sea cabled natural observatory located 200-300 meters below the sea level. One robot has a wide-hourglass-shaped chamber that by turning it upside-down can capture around 20 liters of sediment and suspend it in the seawater column (images of the wider camera angle are also captured for context). A smaller arm on the robot catches the sediment as it falls, while colonizers, such as the enigmatic giant protozoan, Xenophyophore, are temporarily in suspension. After leaving this moving vertical "funnel" for a time, the mudflow can be released to land on the sediment surface while still conducting experiments to see what damage has occurred. Other instruments even allow us to manipulate the sediment at the molecular level, albeit in the quite restricted laboratory environment. Overall, these emerging techniques are decisive in being at the cutting edge of a new phase of biodiversity none of us could have dreamt we would live to see.